IMAGES
of America

CASCADE LOCKS AND CANAL

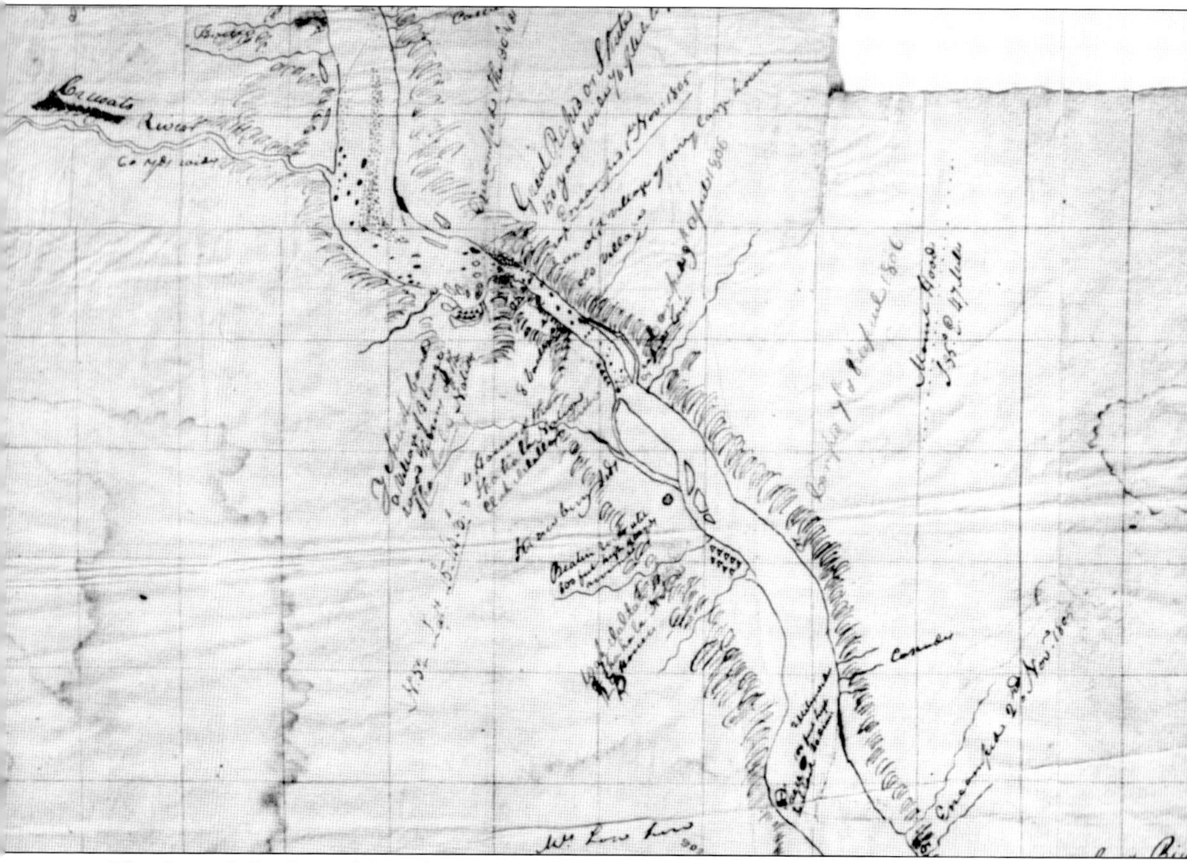

The Cascade Locks and Canal were cut through a low-lying, rocky section of the Columbia River bank. The river constricts to 150 yards wide due to accumulated debris from a series of landslides referred to as the Cascades Landslide Complex. This map from the Corps of Discovery shows the four and a half miles of the Cascade rapids, as well as encampment sites and measurements. (Map by William Clark, courtesy of Oregon Historical Society.)

On the Cover: The *Bailey Gatzert* was one of the fastest ships on the Columbia River and could make the trip from Portland to The Dalles in under five hours. It was a very popular passenger ship that made hundreds of trips and carried thousands of passengers through the Cascade Locks and Canal. (Courtesy of Oregon Historical Society.)

IMAGES
of America

CASCADE LOCKS AND CANAL

Friends of the
Cascade Locks Historical Museum

ARCADIA
PUBLISHING

ISBN 978-1-4671-0816-4

Published by Arcadia Publishing
Charleston, South Carolina

Printed in the United States of America

Library of Congress Control Number: 2022937590

For all general information, please contact Arcadia Publishing:
Telephone 843-853-2070
Fax 843-853-0044
E-mail sales@arcadiapublishing.com
For customer service and orders:
Toll-Free 1-888-313-2665

Visit us on the Internet at www.arcadiapublishing.com

*To the residents of the Cascades of the Columbia,
past, present, and future.*

Contents

ACKNOWLEDGMENTS

Since the Cascade Locks Historical Museum was founded in the early 1960s, dozens of community historians have worked to capture historic moments in our community. The museum would not exist without the continued support of the City of Cascade Locks and the Port of Cascade Locks. We are grateful to the staff and volunteers who stewarded, preserved, and displayed the collection, including Wayne and Lottie Gurley, Lois A. Harrison Williams, Arthur and Elsa Warren, Ben and Bea Scott, Jean McClean, Faye Nelson, Irma Davison, Mary Stillwell, Paulina Updyke, Pat Power, and Brittany Churchwell.

Founded in 2013, the Friends of the Cascade Locks Historical Museum (FCLHM) captured the energy of the community, creating a nonprofit to support and manage the museum. At the time of publication, the board of the FCLHM included Debora Lorang, Ken Royer, John Stipan, Terry Zwanziger, Ralph "Butch" Miller, JoAnn Wittenberg, Dan Yates, and ex-officio members Gordon Zimmerman, city administrator, and Olga Kaganova, port general manager.

This book was compiled by FCLHM executive director Janice Crane. Janice spent many hours poring over a vast library of resources compiled by dedicated community historians to generate the captions for these images, and could not have completed this project without their recorded knowledge. She is particularly grateful for the meticulous records kept by chief lock tender Valentine Tomkins, the detailed compilations of US Army Corps of Engineers historian William F. Willingham, the digitization project team at the Corps of Engineers archives for scanning and optimizing historic reports, and the local knowledge records of Jim Atwell. She is also grateful to her husband, Will Hobson, and Arcadia Publishing editor Caitrin Cunningham for their patience and support throughout the project.

Prints of all of the photographs in this book are in the collection of the Cascade Locks Historical Museum. The museum's photograph collection has been supplemented by donors ordering prints from other organizations and then donating the prints to the museum. While every effort has been made to identify the original institution that provided prints, negatives, originals, and copies of these photographs may also be found in other collections. We have identified the source of photographs found in our collection where possible and indicated them in the courtesy lines.

INTRODUCTION

Nearly 600 years ago, a great earthquake started a series of landslides on the north bank of the Columbia River. The collapse of the faces of Table Mountain and Greenleaf Peak temporarily blocked the whole river. As water washed away debris, the path of the river changed and rushed down a narrower course. The large boulders remaining from the collapse formed the Cascade rapids. The Cascades were four and a half miles in length, just 150 yards wide at their narrowest point, and had a total low water fall of 36 feet.

The Cascade Rapids have been a vital salmon fishery and gathering place for many Native American nations since time immemorial. Portage roads on both banks facilitated trade and travel between coastal and inland communities. Lewis and Clark used these existing portage roads on their journey to the Pacific Ocean. As European fur traders came to the region, they disregarded Native authority over the portage and used violent force to take control of the roads. As violence increased, so did occupation by the US Army. In 1855, treaties were negotiated with the tribes on the north and south banks of the river. The treaties were ratified, and the people were forced to reservations by 1859. Four of the tribes maintain seasonal fishing, hunting, and gathering rights in the area, and Native fishing platforms are still used in the remnants of the Cascade Locks and Canal.

The Cascades posed a challenge for navigating the Columbia River. When European settlers entered the region, they broke the river into segments and kept boats above and below the rapids. As industry and travel boomed, transportation innovation around the Cascades took priority. White entrepreneurs competed for portage tolls on either side of the river. They laid tracks for mule trains on the north bank beginning in 1851 and on the south bank in 1857. After two years of fierce competition, John Ainsworth brought both tramways under the banner of the Oregon Steam Navigation Company. The mules on the south bank were replaced in 1862 by the *Oregon Pony*, the first locomotive built on the west coast. The tramway on the north bank was replaced with steel tracks and a steam locomotive in April 1863. All river traffic was diverted to the north bank, and the *Oregon Pony* was decommissioned.

This monopoly on transportation brought higher prices to freight and settler movement. When the transcontinental railroad arrived, it reduced the region's dependence on water routes, making freight rates more competitive. Shippers and steamboat lines demanded that the US Congress make improvements along the Columbia River. The Army Corps of Engineers began studies for the canal in 1874. The result of the study was a proposal to build a canal at the Upper Cascades and to remove the most dangerous rock and debris in the Lower Cascades. Congress allocated $90,000 in 1876 for the final planning and to begin construction on the canal.

The engineering and construction of the locks took over 18 years. The project began in 1878 and was not finished until 1896. Construction was slow and difficult. Seasonal flooding meant that workers could only access the site during the fall and winter. The federal government sought to purchase the land to build the canal from the Oregon Steam Navigation Company (OSN), which quoted a price of $200 per acre with access addendums. After months of unproductive negotiations, the government determined to condemn the land rather than purchase it. A jury awarded OSN $7,500 for the plot.

The first contract was awarded to Ball and Platt, a construction firm from New York. Only two months into the project, A.H. Ball, a senior partner and project supervisor, died of exposure to the winter elements while supervising construction. After his death, there was a lack of leadership, hazardous weather, a legal battle between the company and his heirs, poor living conditions, and no nearby medical facilities. Workers struck in 1879, delaying construction further. One hundred white laborers protested for higher wages and against the employment of the hundred Chinese workers on the project. At the time of the strike, Chinese laborers were paid 85¢ per day and white laborers were paid $1.75 per day.

In 1880, the Corps of Engineers took over direct supervision of the project. Under the direction of Willard Young, the government built better housing for workers, supplied room and board, and hired a doctor. Around 350 construction workers stayed in barracks on the federal property during peak construction season.

In 1892, the project was turned back over to a contractor, the J.G. & I.N. Day Company of San Francisco. In 1894, a record-breaking flood delayed the construction, damaging bulkheads and some of the existing groundwork. While the locks were finally functional in 1896, finishing touches continued through 1914.

The locks at the Cascades finally opened on November 5, 1896. Parties took place at the locks and in Portland, The Dalles, Vancouver, and Camas. Hundreds of people passed through the locks on steamboats to view the final project. The steamer *Sarah Dixon* had a small cannon on deck and blasted salutes. Through the locks also went *Regulator, Dalles City, Maria, R.R. Thompson, Water Witch*, and *Harvest Queen*. *Sarah Dixon* and *Regulator* raced to be the first boat to The Dalles. *Sarah Dixon* won, arriving a full 30 minutes ahead of *Regulator.*

The benefits of the locks and canal to river commerce were immediate. In the first six months, 4,941 tons of freight and 9,081 passengers passed through the canal. That number more than tripled in 1898 to 18,812 tons of freight and 30,327 passengers. The value of freight through the locks in most years exceeded the cost of construction, and portage freight charges were eliminated. This restored competition between the railroad and multiple river shippers. The construction of the Cascade Locks and Canal provided valuable experience for the next project of a canal and locks at The Dalles, and all future Columbia River projects.

The trip from Portland to the Dalles became an easy day trip with the navigational improvements. Scenic cruises through the Gorge increased in popularity as the journey became less difficult and ships were built for speed. Tens of thousands of passengers passed through the Cascade Locks each year.

The Cascade Locks were formally decommissioned when the Bonneville Dam was completed in 1938. The new dam raised the low water level by 12 feet at the Cascade Locks, and the new water depth covered the lower lock and submerged the Upper Cascades. The upper lock gates were removed in 1953. The canal, still visible in Cascade Locks Marine Park, is now used for recreational fishing.

One

The Cascade Rapids and Pre-Lock Portage

The locks at the Cascades were built to bypass the Upper Cascades section of the Columbia. The whole volume of the Columbia River gushed over rocky reefs and landslide debris, making a spectacular show of churning white water completely impassable by boat at low water and inadvisable at high water. (Courtesy of Oregon Historical Society.)

The Cascades earned their name at high water. Prior to the construction of dams, the Columbia River had a high to low water range of up to 60 feet. These two images show the same rock formations in the Upper Cascades at different times of the year. The image above shows the rocks above low water in the fall, and the image below shows the flow of the seasonal June floods. The south face of Greenleaf Peak, visible below, bears the scarring and exposed cliffs that remained after one of the landslides that was the source of these obstructions. (Both, courtesy of Oregon Historical Society.)

The Cascade rapids were a vital salmon fishery. Fish swimming upstream to spawn could be captured efficiently due to the bottleneck of the river, the shallows, and flat access points created by the landslides. The Columbia River ecosystem supports 61 species of fish. Enormous sturgeons lurk in the deep eddies, while ancient lampreys, meaty chinook, sleek coho, and succulent sockeye salmon and steelhead trout fight the currents on their journey inland. Native peoples build wooden platforms to access the fast-moving water with dip nets and gillnets. Fish would return from the ocean to the Columbia River in the tens of millions. Euro-American settlers brought new technology and preservation techniques to harvest and distribute what would be unsustainable quantities of these fish. The structures on the left are fish wheels used by the Warren Cannery. (Courtesy of Oregon Historical Society.)

This is a closer look at the basalt boulders in the Upper Cascades. The series of landslides is known as the Cascade Landslide Complex. Between 600 and 250 years ago, the most recent of these landslides covered five and a half square miles of the river basin, burying the previous Columbia River channel and diverting the river flow toward the Oregon side. It dropped sediment and boulders as large as 800 feet long and 200 feet wide, which temporarily blocked the river with an earthen dam up to 200 feet high. This blockage became known as the legendary "Bridge of the Gods," a natural river crossing. The exposed cliff faces on Table Mountain, Greenleaf Peak, and the Red Bluffs still show where the land fell from the mountains to the river. These boulders remained after the formation collapsed and washed away, making up the sections of the Cascade rapids. (Photograph by Carleton E. Watkins, courtesy of Oregon Historical Society.)

These two images of the Upper Cascades were taken from approximately the same angle, one during springtime high water (above) and the other at low water. During seasonal flooding, the rocks and debris in the Upper Cascades would be completely submerged, allowing for daring captains to attempt to run the rapids by ship. These rocks broke the surface during low water, making this section of the river completely impassable most of the year. The portage railroad trestle in the background was built to accommodate the seasonal flooding, barely staying above the crest of the river most years. The docking wharf boat floated alongside the rail trestle for moving people and freight. (Above, courtesy of Oregon Historical Society; below, courtesy of Cascade Locks Historical Museum.)

The Corps of Discovery reached the Cascades in October 1805 and returned past them in April 1806. Upon reaching the rapids, the Corps took several trips along Native portage paths, determining that the rapids were six miles long and that all of their belongings must be ported, not sent over the chutes. On October 31, 1805, William Clark described the Upper Cascades as a "Great Shute or falls about 1/2 a mile with the water of this great river Compressed within the Space of 150 paces in which there is great numbers of both large and small rocks, water passing with great velocity forming & boiling in a most horriable [*sic*] manner, with a fall of about 20 feet, below it widens to about 200 paces and current gentle for a Short distance." (Both, courtesy of Cascade Locks Historical Museum.)

Massive boulders left over from landslides littered the river. These rocks that were near the mouth of Ruckel Creek made rapids in the Middle Cascades. These boulders and many like them were blasted with dynamite during the construction of the Cascade Locks to clear the lower six miles of river for access to the locks. (Courtesy of Cascade Locks Historical Museum.)

Humans and freight needed portage around the Cascades. Early settlers took advantage of existing Indigenous footpaths to create tolled roads. With the promise of donation land claims and the discovery of gold came a rapid increase in river traffic and demand for ship and wagon routes. Ship traffic, portage railroads, and fish wheels popped up along the river quickly as Euro-American settlers staked claims to natural resources. (Courtesy of Cascade Locks Historical Museum.)

The people who built early fish wheels determined the location of permanent structures based on observing where Native fishers were successful with dip nets. The image above shows one of Frank Warren's fish wheels, a fixed structure with a log foundation and set height. The image below shows a scow fish wheel in the Middle Cascades. Scow fish wheels were a later development that could rise and fall with the level of the river, move, and operate year-round. The quick flow of the shallow water over the rapids spun the wheels and directed thousands of tons of fish into the path of the wheels. (Both photographs by Carleton E. Watkins, courtesy of Cascade Locks Historical Museum.)

The first railroad in the Washington Territory and in the Pacific Northwest was a wooden rail tramway constructed by Francis Chenoweth in 1851. Chenoweth moved to the Cascades from Wisconsin in 1849, staking a donation land claim along a sandy delta at the base of what would come to be known as Hamilton Mountain in North Bonneville, Washington. The initial tramway he constructed was just two and a half miles long, meeting with the portage path and passing the Lower Cascades that bordered his land claim. The tramway operated with mules pulling crude, wood-hewn flatcars over fixed wooden rails. Chenoweth charged exhausted emigrants for passage to use this reprieve from the trail and made a decent amount of money portaging their goods around the rapids. Chenoweth sold the business to brothers Daniel and Putnam Bradford in 1853. (Photograph by Carleton E. Watkins, courtesy of Cascade Locks Historical Museum.)

The Bradfords made improvements to the Washington Territory portage. They lengthened the wooden track to four miles and joined it to their store and landing at the Upper Cascades. This landing was difficult to navigate and required ships to pass through a narrow channel between the shore and a rocky island to meet with the wharf boat. In 1855, the United States broke its treaty with the Yakama Nation. The treaty stated that the tribes had two years to move from their ceded lands to the reservation, but the United States opened the lands to settlement only 12 days after the treaty was signed. This led to a series of escalating attacks that came to be known as the Yakima War. The Bradfords' store was the site of a raid, disrupting service on the portage railroad. Reciprocal violence prompted the military to explore river improvements and drove emigrant traffic exclusively to the Oregon shore of the Columbia for a short time. (Photograph by Carleton E. Watkins, courtesy of Cascade Locks Historical Museum.)

In 1856, Joseph Ruckel and Harrison Olmstead built a tolled mule-and-wagon portage road on the Oregon side of the rapids. Emigrants now had to pay for safe passage on the Oregon side of the river as well. The road followed the Native footpath used by Lewis and Clark. This tolled road was the beginning of fierce competition for tolls on the Columbia River. Ruckel negotiated contracts with the government to move freight over his road, cutting deeply into the Bradfords' profits. In 1857, Ruckel began the process of converting the road into a wooden tramway but abandoned the project in 1859. In 1861, the Oregon Portage Company worked with Ruckel and Olmstead to finish a wooden tramway along the toll road. Another set of mule-pulled carts on wooden tracks, the track was built at a broad gauge of five feet. (Photograph by Carleton E. Watkins, courtesy of Cascade Locks Historical Museum.)

Photography was not prevalent in the Northwest while emigrants were dragging their belongings along crude wooden railroads. This photograph of a portage railroad was posed for and taken by Carleton E. Watkins in 1867, four years after this stretch of the railroad was purchased and closed in favor of a Washington Territory upgrade. One of the goals of the photography series was to emulate the experience of using the railroad and to capture their historic importance before the wooden tracks were replaced with steel rails. In this photograph, Joseph Bailey (left) and an unidentified man pose on an empty mule cart. Watkins's camping gear is visible on the right. (Photograph by Carleton E. Watkins, courtesy of Cascade Locks Historical Museum.)

Joseph Bailey was the superintendent of the railroad and sawmill when OSN took over the management of the Oregon Portage Railroad in 1859. He lived on-site at the sawmill with his wife and children. This sawmill was at the mouth of Eagle Creek, near where the Bonneville Dam is today. Thousands of board feet of six-inch-square fir were strapped with iron and used to build the center of the tracks and rail for the carts. The space between the rail ties was covered with oak planks to make a walking path in the right-of-way. The banks along the Columbia River are prone to flooding, and the sawmill was kept busy replacing rail sections, bridges, and trestles throughout the brief three-year operation of the railroad. (Courtesy of Oregon Historical Society. Photograph by Carleton E. Watkins.)

Wooden trestles were easier to build than hewing flatter paths in the rock along the steep portage road. The Oregon Portage Railroad was susceptible to flooding, and trestles were replaced regularly. This trestle bridge was around Tooth Rock, between the confluence of Eagle Creek and the Columbia River. A man is driving a mule cart in an 1867 demonstration. (Photograph by Carleton E. Watkins, courtesy of Cascade Locks Historical Museum.)

The Oregon Portage Railroad was the first railroad in Oregon. The lower landing of the railroad was the ship access point at the mouth of Tanner Creek. It was sometimes called "Middle Landing" because it was the entrance to the middle section of the Columbia. Ships could dock to unload freight for portage upriver and pick up passengers to take downriver. (Photograph by Carleton E. Watkins, courtesy of Cascade Locks Historical Museum.)

In 1859, John Ainsworth incorporated both railroads into the Oregon Steam Navigation Company. Both portages operated independently, but the rates for shipping and passengers were set, and profits were shared. In 1861, the Bradfords announced plans to transform their portage into a steam railroad with iron rails. Ruckel announced a similar plan for the Oregon portage but using the existing wooden tracks. (Courtesy of Oregon Historical Society. Photograph by Carleton E. Watkins.)

Running the "impassable" rapids became a challenging feat for daring ship captains. Stern-wheeler captains would attempt to best each other, running the rapids at lower and lower water levels, occasionally losing their ships. In this photograph, many spectators lined up on the Washington shore to watch Capt. J.W. Troup run the *Hassalo* through the Cascades. (Photograph by Orlando Hofsteater, courtesy of Oregon Historical Society.)

The *Oregon Pony* was commissioned for the Oregon Portage Railroad in 1861. This was the first complete locomotive built on the west coast. It was made by Vulcan Iron Foundry in San Francisco. Vulcan Iron Foundry primarily made mining equipment and assembled locomotives that were shipped in pieces. The *Oregon Pony* was built to run on the specifications of Ruckel's wooden track and wide gauge. While recognized even at the time as a historic achievement, the *Oregon Pony* only operated on the Oregon Portage Railroad for one year. From April 1862 to April 1863, it hauled 200 tons of freight per day upriver, and hundreds of pounds of gold dust and travelers downriver. Freight was unloaded from Lower Columbia ships onto the docks at the lower landing, piled onto flatcars, and chugged along by the *Oregon Pony* to the ships on the Middle Columbia. (Courtesy of Cascade Locks Historical Museum.)

The *Oregon Pony* was decommissioned by OSN shortly after iron rails were completed on the Washington Portage in 1863. It was shipped upriver to the next portage at The Dalles. After sitting unused for three years, it was sold to the Steam Paddy Company and landed back in San Francisco, performing street improvements until its retirement in 1873. After decades dormant in a warehouse, the *Oregon Pony* burned in a fire in 1904, losing its brass bearings, wooden frame, and canopy roof. An employee of Steam Paddy Company recognized its historical significance and ensured it was restored in time for display at the Lewis and Clark Exposition in Portland in 1905. The *Oregon Pony* remained on display at Union Station in Portland for many years. It was brought back to Cascade Locks in 1970 and restored again in 1980 to its current state. (Courtesy of Oregon Historical Society.)

OSN developed a monopoly on all trade and travel on the Columbia River. The Corps of Engineers began a study for locks and a canal in 1874. In his 1876 report, district engineer Maj. John M. Wilson wrote, "It is evident that the interests of this great and rapidly-developing portion of our country will demand the construction of canals around these obstacles, now passed by railway portage," and "the absolute necessity for such improvements" open "the route via the Columbia," which is "the only one by which the products [of Oregon, Washington, and Idaho] . . . can be brought into market." In his 1878 report, Wilson reiterated the urgency of disbanding the monopoly. "The whole of the navigation of the Upper Columbia and Snake is now in the hands of a single company. It owns and operates all the steamers on the waters . . . and has railroad portages around the Cascades and between The Dalles and Celilo. All freight has to be handled twice at each of these places." (Photograph by Carleton E. Watkins, courtesy of Cascade Locks Historical Museum.)

Two

Construction of the Locks and Canal

Building the canal and locks began in 1878. Construction was slow and difficult. Seasonal flooding meant that workers could only access the site from late fall through early spring, when working conditions were poor. This early photograph shows the beginnings of framing outbuildings and excavation equipment, with a ship passing in the background. (Courtesy of Cascade Locks Historical Museum.)

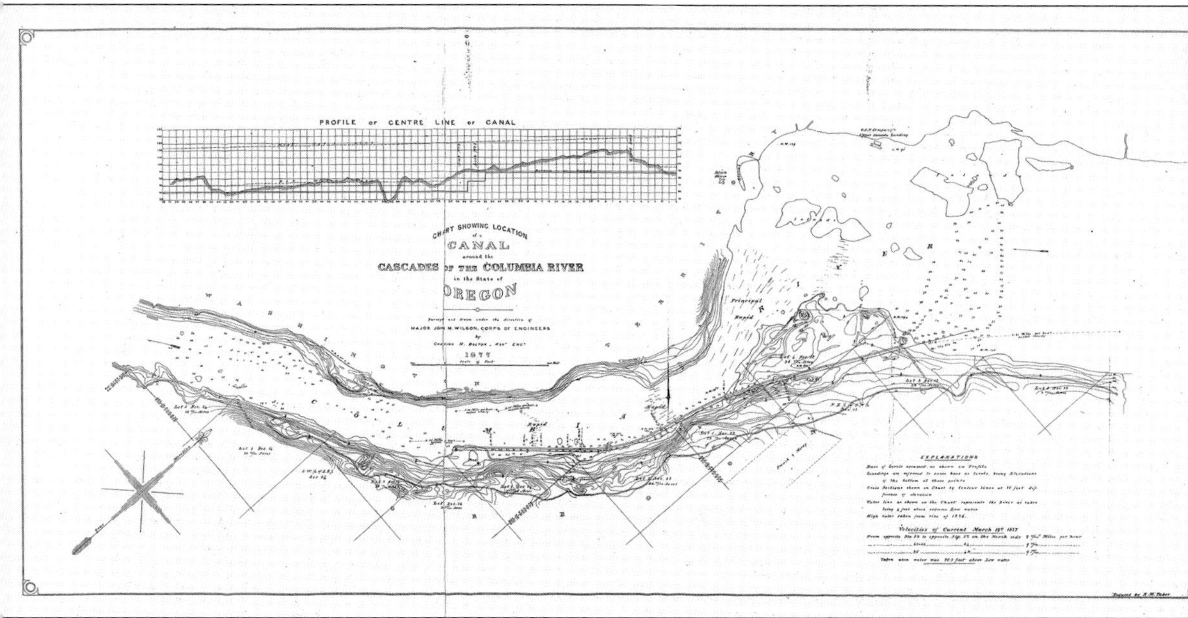

The desired land for the canal was owned by Oregon Steam Navigation Company, which offered to sell it in 1875 for $200 per acre (approximately $9,500 for the plot), reserving right-of-way for a railroad. The secretary of war declined the offer, instead opting to proceed with condemning the land. After lengthy, unproductive negotiations, murkiness around the land title, and the condemnation procedure, a jury awarded $7,500 to OSN for the property. In 1877, the Corps of Engineers hired assistant engineer Channing M. Bolton for the Cascades Canal project based on his 17 years of experience surveying and building specialized locks and canals in the eastern United States. Unsatisfied with the 1874 survey work, Bolton took a team and completed his own survey of the area, drawing this map and indicating that the locks should be constructed through existing low-lying ponds. The updated estimated cost for the project was $1.7 million. (Courtesy of US Army Corps of Engineers.)

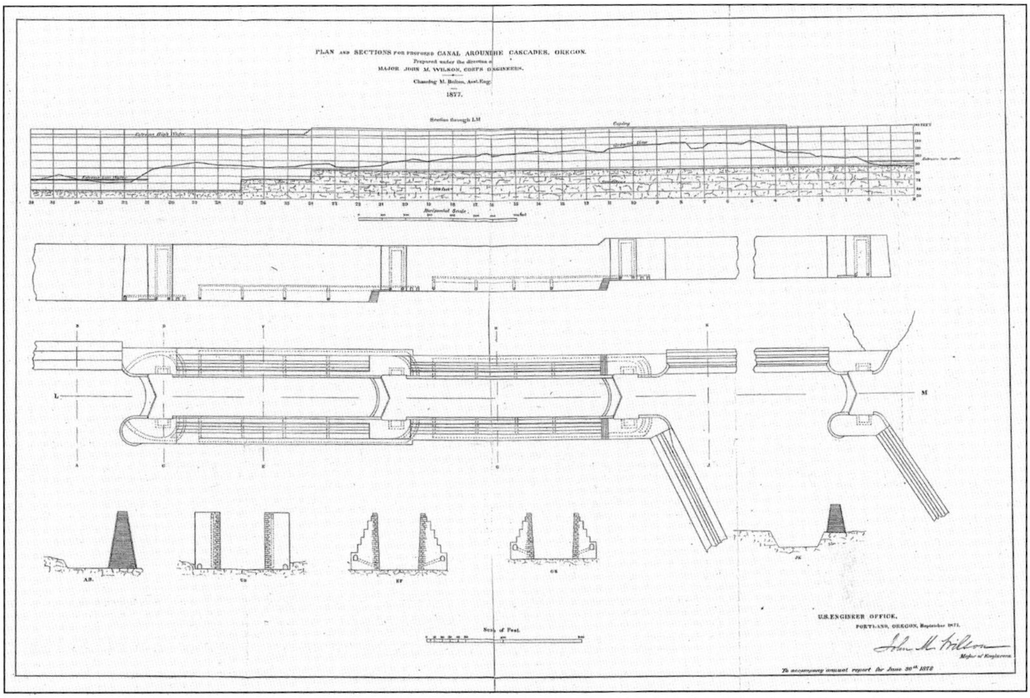

Bolton designed the canal to be 7,200 feet long, 50 feet wide, and eight feet deep. It featured two locks with 12-foot and 14-foot lifts, which would be 300 feet long and 70 feet wide. The locks were to be filled and emptied using culverts in the side walls and would open using turbine and miter wheels. (Courtesy of US Army Corps of Engineers.)

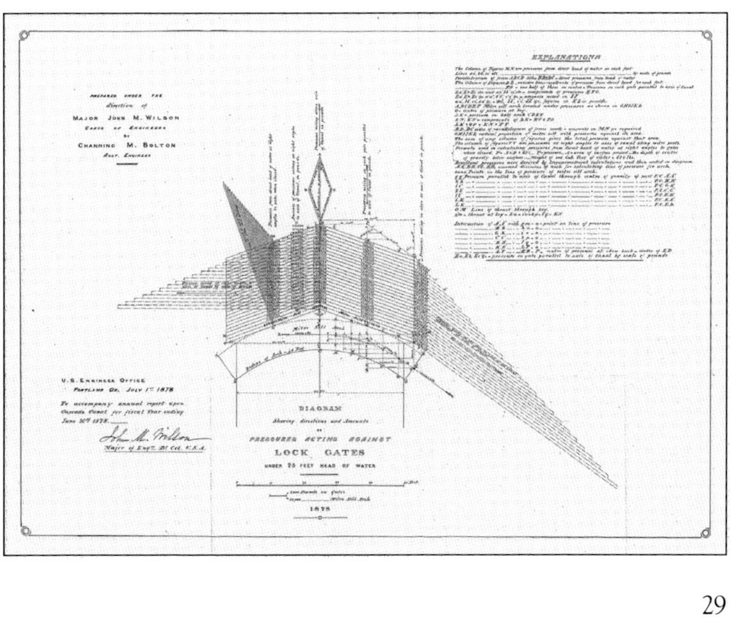

Bolton took it upon himself to build a scale model of his design to demonstrate how his hydraulic power mechanisms would operate the proposed wooden lock gates. The guard gates were to be constructed of wood and be 54 feet high each. This diagram shows the pressure that the river would put on the lock gates at 25 feet of water. (Courtesy of US Army Corps of Engineers.)

The initial contract for construction was awarded to Ball and Platt of New York. The first blast broke ground on October 30, 1878. A.H. Ball, senior partner and project manager, died of exposure at the project site on January 4, 1879. Between his death and a resulting lack of management while the estate navigated succession and multiple strikes, very little was accomplished during the two-year contract. During the first two seasons, 30 to 70 men worked removing earth by hand and beginning a rock quarry near the construction site. White and Chinese laborers worked under this contract, and racial tension was a factor in the strikes. Working and living conditions continued to deteriorate as ice, snow, and rain isolated the workforce at the Cascades. Poor housing and lack of decent medical care lowered morale. Bolton reported that after one accident, he personally used a sewing needle and silk thread to stitch up a laborer's wounds. Bolton resigned from the project in May 1879. (Courtesy of Cascade Locks Historical Museum.)

The Ball and Platt contract was terminated on November 20, 1879. At this time, the Corps of Engineers purchased the equipment on-site and took over management of the project. Under Corps management, new worker housing was constructed and a doctor was hired to improve working conditions. The new local-in-charge, Lt. G.F. Powell, reported that "the men are now comfortably lodged, and excellent meals are provided them . . . at the cheap rate of $3.75 per week." Most of the work under government management was completed by hired laborers, managed by civilian engineers or Corps officers. By 1880, the town of Cascade Locks consisted of 100 permanent residents and 350 transient laborers. (Both, courtesy of US Army Corps of Engineers.)

Over the 18 years that it took to complete construction, 817,825 cubic yards of earth and rock were excavated to create the canal. The upper layer of sediment was a combination of silt, sand, pebbles, and boulders from the landslide and river deposits. The bottom of the lift chambers in the locks was blasted and carved from the solid basalt riverbed. Most of the earth and stone were removed using hand tools, then tracked out using mule-drawn carts. The bedrock disintegrated easily after it was exposed to air and water, and had to be covered with a thick layer of concrete. (Both, courtesy of US Army Corps of Engineers.)

Earthen embankments were used in early construction to hold back the water while lock pits were dug and canal walls were formed. The calm waters directly above the rapids and canal entrance were home to the first ship-building yard on the Middle Columbia. The Attwell boatyard was the builder for the steamships that traveled between the Cascades and The Dalles. (Courtesy of US Army Corps of Engineers.)

As the canal was cut through the land, debris was used to build up the embankment. Riprap walls were constructed to prevent erosion. Quarrying stone and building up the embankment were tasks that could be completed during the rainy winters and annual floods at the construction site. (Courtesy of US Army Corps of Engineers.)

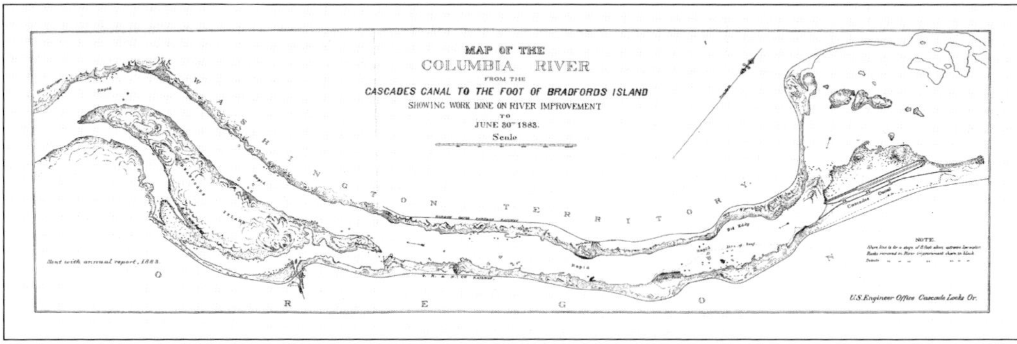

During the 1880 season, Lt. G.F. Powell noted that the low water level was two and a half feet lower than calculated for the original canal design. This meant that for fully-loaded ships to use the canal, improvements needed to be made in the approach through the Lower and Middle Cascades. The canal was deeper than the river in many places, and channels needed to be blasted through five miles of rocks. Work at the canal site essentially halted from 1881 to 1886 while these improvements were made. The locations marked in black on the map above show where blasting occurred in 1883. The photograph at left shows an explosion to clear the channel. (Above, courtesy of US Army Corps of Engineers; left, photograph by Carleton E. Watkins, courtesy of Oregon Historical Society.)

Over the next five years, great quantities of rock and reef were removed to clear access to the canal. The project was first delegated to private contractors, but after unsatisfactory work, the Corps of Engineers resumed direct management of the project. Many of these large boulders were blasted with dynamite. Seven workers lost their lives during the rock removal process. (Photograph by Carleton E. Watkins, courtesy of Oregon Historical Society.)

On February 16, 1884, the water level again reached two and a half feet below extreme low water. Several ships participated in test runs to help identify where additional improvements to approach the canal were needed. The *Lurline* made it to Ruckel's landing, *McCully* failed at the Middle Blockhouse, and the 80-foot scow *Snowflake* made it to the Big Eddy Rapid, just below the canal entrance. (Courtesy of Cascade Locks Historical Museum.)

After losing and restarting his award-winning photography business, Carleton E. Watkins returned to the Cascades in 1882 to complete a new series of scenic photographs. During this trip, he captured scenes of early construction on the Cascades Canal. The photograph above shows the construction of a wooden barricade to keep water from backflowing into the excavation, with the Cascade Rapids and portage railroad as a scenic backdrop. The photograph below shows progress inside the canal. Derricks were used lifted heavy stones and supplies between the construction area and the mule-pulled rail carts. (Both photographs by Carleton E. Watkins, courtesy of Oregon Historical Society.)

Construction in the canal bed was only one-third complete when work paused for river improvements. Construction resumed in earnest in 1886 after the lower river improvements were complete. The workers continued to excavate the canal bed, and stonemasons were brought in to cut to the precise measurements needed to build the walls. The lowest level of excavation had to be completed without explosives so the bedrock would be stable enough to support the masonry. In this photograph, two teams of workers use shovels to load sediment from between broken boulders into carts adjacent to the canal walls for removal. Excavation of the canal bed was the most labor-intensive process and continued right up until the canal opening. Stone excavated from the canal floor was faced for paving the embankment above the canal walls. (Courtesy of US Army Corps of Engineers.)

Earthen embankments were built up seasonally to provide access to the canal site. During the late spring, water levels would rise and the protective barriers would usually fail and allow some seepage to flood the canal. The image above shows men building up the embankment and the drained canal floor site covered with river deposits. The image below shows the failed embankment and flooded work site. Capt. Philip Price, a Corps engineer in 1883, reported that "the great rise of water during floods causes a very large annual expense for the moving and protection of this . . . property. It must be all either moved or effectually protected at the beginning of the flood, and must be again moved into position and put in readiness for work on the subsidence of the water." (Both, courtesy of US Army Corps of Engineers.)

The geology of the bedrock below the Columbia River was largely unstudied before the Cascades Canal project. Core samples were bored to assess the process over the course of the project, showing layers of sand, clay, gravel, shale, basalt boulders, soft and hard slate, and at the deepest level, a very hard sandstone. Above, men use shovels and picks to remove earth and rocks from the lower entrance to the canal. The carts they use to move the debris follow short tracks to dump just past the end of the canal entrance. Once workers reached the bedrock at the lowest part of the canal, the work had to be done by hand. (Both, courtesy of US Army Corps of Engineers.)

In 1879, Oregon Railway and Navigation Company (OR&N) purchased Oregon Steam Navigation Company and Oregon Steamship Company. The monopoly along the Columbia River was thus expanded and strengthened by OR&N. Freight rates along the portage in 1886 were up to 10 times the rates of portage rails in other parts of the country. In 1891, the State of Oregon built

its own portage railroad on trestles around the canal worksite, parallel to the OR&N rail line. With the river improvements, ships could dock at a wharf boat just below the canal site. In this photograph, the *Dalles City* waits for a transfer on the state-owned portage rail, while the OR&N line sits empty. (Photograph by Herbert A. Hale, courtesy of Oregon Historical Society.)

On January 24, 1893, J.G. and I.N. Day Company was contracted to complete the canal project. Construction at the lower end of the canal required bulkheads to keep backwater from flooding the worksite. These bulkheads were left in place over the high-water season and any water that came in during the annual freshet would be pumped out. (Courtesy of US Army Corps of Engineers.)

The paved banks were supported while they settled and dried using massive timber beams. This photograph shows the angle of a base support beam and how it was set on a platform with boulders to maintain appropriate pressure on the freshly paved banks. (Courtesy of US Army Corps of Engineers.)

The walls of the locks required a large workforce of skilled stonemasons. They faced 85,491 cubic yards of blocks in the stonecutting yard adjacent to the canal work site. Facing the blocks was precise work, requiring careful craftsmanship to create flat, even surfaces and consistent sizes and lengths. The blocks were faced entirely by hand using chisels and gads. Teams were hired from around the world, bringing in experienced workers from Italy, Sweden, and the eastern United States. This crew was from Vermont. From left to right are (first row) Julius J. Carlson, Andy Vanstrom, Carl Carlson, Otto Schmidt (with dog), Thad Glazer, Charley Gray, Dave Merrill, and Lawrence Peterson; (second row) George Glazer (standing), Merl Yettick, Swan Peterson, eight unidentified, and J.F. Hendrick (standing, far right). (Courtesy of History Museum of Hood River County.)

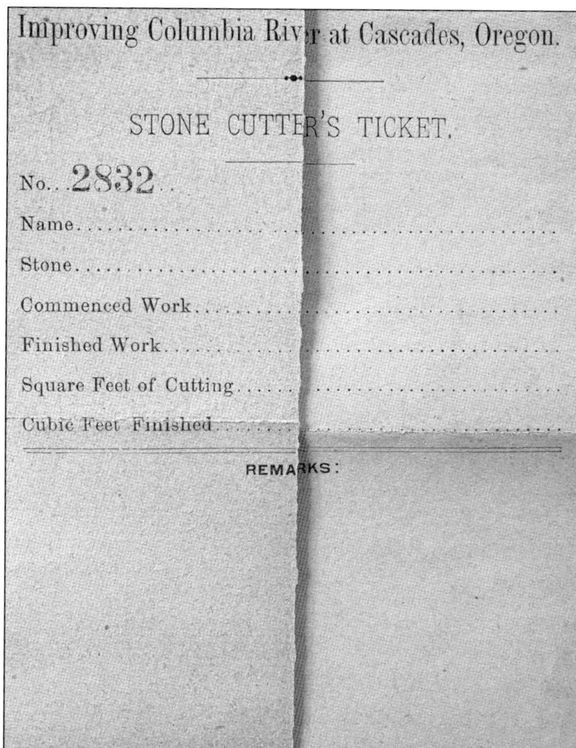
Improving Columbia River at Cascades, Oregon.

STONE CUTTER'S TICKET.

No. 2832

Name..........

Stone..........

Commenced Work..........

Finished Work..........

Square Feet of Cutting..........

Cubic Feet Finished..........

REMARKS:

Most of the stone used in the walls was basalt sourced from the worksite or quarried near Herman Creek, Oregon, and Washougal, Washington. A single cubic foot of this type of stone weighs 176 pounds. A total of 85,491 cubic yards of stone masonry was used in the locks, with an additional 117,852 cubic yards used to pave the surrounding banks. (Courtesy of Oregon Historical Society.)

The stonecutter's tickets were a way of measuring labor and material used. Workers would be assigned a card and number, and the manager would measure and tally the stone they cut. This recording system is the reason such exact records are available for the quantity of stone used in the canal. (Courtesy of Cascade Locks Historical Museum.)

About 1,400 feet of culverts along the bottom of the canal allowed water to flow or drain to raise or lower ships. The filling culverts had 11 openings in the walls of the locks, each three feet by five feet. The emptying culverts were below the lower guard gate and had openings nine feet wide by 10.5 feet high. The turbine culvert was also five feet by three feet. The valves to open and close the culverts were made of bronze. Granite was brought in from Spokane, Washington, and Fresno, California, to line the culverts. (Both, courtesy of US Army Corps of Engineers.)

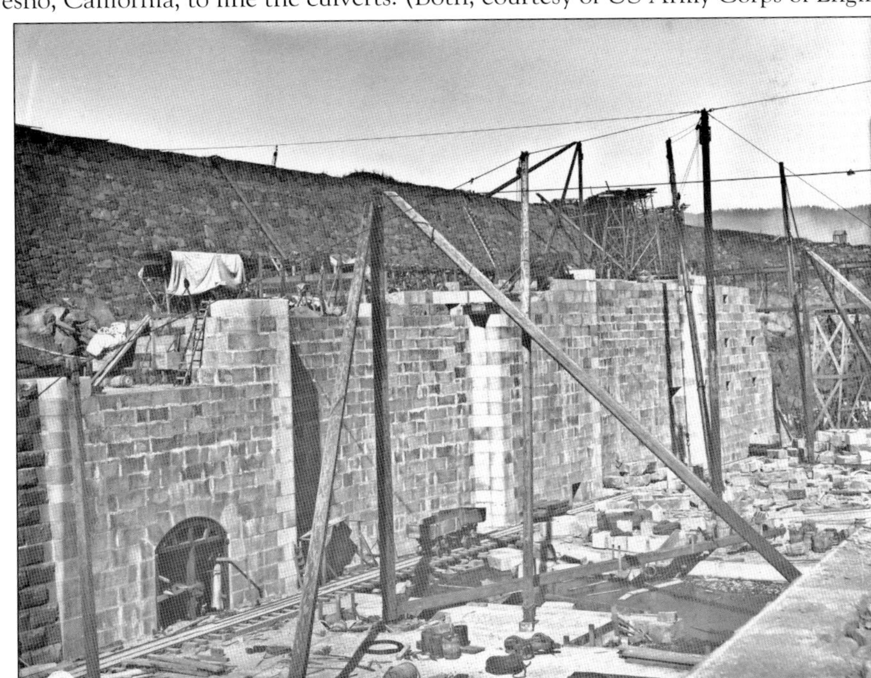

The lowest level of bedrock in the canal was sandstone. Because it disintegrated easily once exposed to the damp atmosphere, the bottom of the canal was covered with a thick layer of cement. The engineers sought high-quality materials for the project and had an on-site testing facility for cement. They would make small cubes of cement and subject them to different pressures and conditions to determine which brand would be most appropriate for long-term use. Only Portland cement met the rigorous quality standards for the project. The walls rested on the concrete-covered bedrock. The blocks for the walls were layered and also mortared with this cement. The banks above the canal walls were then protected from high water by adding three-foot-thick basalt blocks. (Courtesy of US Army Corps of Engineers.)

From June 4 to 8, 1894, the Columbia River reached its highest stage on record. On June 6, the water measured 145.7 feet at the head of the canal, breaking the 1876 record by six feet. Lt. Harry Taylor was in charge during the 1894 flood. According to the 1894 Corps chief's report, the upper bulkhead "was raised and kept above water level thus preventing the river breaking through and doing great damage to the work already finished." In the photograph above, workers are attempting to prevent the river from washing away the upper bulkhead. Below, "water poured over the protection wall on the north side and finally tore down the lower end . . . 16 to 18 feet of the top of the wall is entirely gone." (Both, courtesy of Oregon Historical Society.)

On June 5, 1894, the water rose to the level of the powder house at the highest point on the northern canal wall. The river completely carried away the earth fill to a point within 90 feet of the top of the canal, leaving a strip only 70 feet wide. The fill had previously been 350 feet wide. Above, workers build up embankments around the tracks to keep the river from flooding over the north side into the canal. The water began to recede on June 8, 1894, as seen below. (Both, courtesy of Oregon Historical Society.)

The floodwaters of June 1894 reached the residences of the chief engineer and local-in-charge. This photograph shows Quarters One, the construction headquarters for Corps of Engineers officers, with floodwater cresting the railroad and flooding three feet into the first floor of the house. The house was later converted into the chief lock tender's residence, not rebuilt after it dried, and left with water damage for the occupants to tolerate. Alice Tomkins Fee, daughter of chief lock tender Valentine Tomkins, lived in the house from 1907 to 1934 and recounted that during her childhood, "in the dining room especially, the floors were warped. So, the table couldn't have casters; it had to sit up on blocks. And we had to remember never to put a guest on a certain side or they'd go over backwards." (Courtesy of Oregon Historical Society.)

Once the floodwaters receded in the fall of 1894, construction and repairs once again began in earnest. Fresh sediment deposits needed to be removed, and the outer walls shored up against rushing river damage. The image above shows the water level at average high water. Equipment needed to be moved annually for the spring freshet and replaced after the site was drained in the fall. The image below shows the damaged canal entrance, sediment deposits, and the remains of a scow fish wheel that washed up on the worksite along the lower canal approach after the 1894 flood. (Both, courtesy of US Army Corps of Engineers.)

Because it washed away significant sections of the riverside of the canal, the 1894 flood prompted adjustments to the plan to include supporting the outer side of the canal wall against erosion. Surveyors were contracted and sent to determine how to best preserve the northern bank. The surveyors measured and evaluated the erosion from the flood and engineers sampled to make sure that the geology was similar to samples from inside the canal work area. The decision was made to pave the riverside of the canal with the un-mortared rubblework that was originally planned to be used on the banks above the canal walls. Lt. Harry Taylor promised the impatient shipping companies and the US Congress that the planned changes would not "delay the opening of the canal a day." (Courtesy of US Army Corps of Engineers.)

The earthen embankments above the canal and on the riverside of the canal were faced with rubble. Initially, extra blocks of un-mortared basalt were stacked and fitted together to protect the form of the banks, and extracted earth and rocks were deposited on the north side to protect the outer wall. The 1894 flood washed away 280 feet of this expansive dirt pile and damaged the un-mortared sections of bank above the canal walls. This highlighted the need to improve the embankments on both the inner and outer walls. The embankments inside the canal walls were mortared to provide additional support. (Both, courtesy of US Army Corps of Engineers.)

After the 1894 updates to the canal plan, faced stones and a layer of cement were added to the newly formed island to secure the banks against future flooding and erosion. Mortared stone was set along the canal walls, and un-mortared rubblestone was placed around the riverside of the island. The image below shows workers placing the new dressed stone above the completed canal wall once the steel gates were installed, and coating it with cement. Ultimately, 117,852 cubic yards of pavement were used on the slope. (Above, courtesy of Oregon Historical Society; below, courtesy of US Army Corps of Engineers.)

In addition to the crew barracks, the Corps of Engineers constructed a building for cement testing and patterns, a warehouse at the landing pier, a wagon shed, and a blacksmith shop. Previously known as Falls, Portageville, and Cascades, the community of Cascade Locks was renamed again and grew as a result of the project. In 1880, the town consisted of about 100 permanent residents and up to 350 construction workers. By 1900, it had grown to 550 permanent residents. In addition to the locks project, the community was supported by a booming fishing industry with multiple canneries and fish wheels in the immediate area. The scenic beauty of the Columbia River Gorge brought tourists from Portland seeking fresh air, spectacular views, and outdoor adventure. Visitors at this time arrived by ship or rail, serviced by the Oregon Railroad (previously Railway) and Navigation Company. (Courtesy of Cascade Locks Historical Museum.)

Bolton's plan for wooden gates was revised after finding that the size of lumber needed to create a sufficient gate was scarce and of questionable durability. The 1888 revisions included upgrading the lock gates to steel. These were produced in Maryland and shipped via rail to the worksite in pieces. The sections were riveted and welded together. Above, scaffolding provides workers access to the gates for riveting. Canal excavation took place right up until the canal opened, as seen here with workers hauling earth on a mule cart while the gates are installed. Below, workers stand on platforms to access the weld spots. At center, one worker can be seen climbing the gate instead of using a ladder. (Both, courtesy of US Army Corps of Engineers.)

At 55 feet tall and 52.6 feet in length, the upper guard gates were the largest steel lock gates built at the time. They weighed 325,353 pounds when fully assembled. As an additional precaution considering the foundation bedrock, the gates were set to overlap with the miter to reduce pressure on the base of the gates. The steel gates were designed to have a tensile strength of 70,000 to 80,000 pounds per square inch to hold back the mighty river. The gates had to be placed and riveted carefully to achieve a tight seal. They were still operated with a hydraulic system, piping water in from nearby Dry Creek (formerly Atwell Creek) to apply pressure to the valves to allow the gates to open and close. The initial wooden pipes were quickly replaced with steel. (Courtesy of US Army Corps of Engineers.)

Millions of board feet of lumber were used in temporary structures to build the Cascade Locks. Due to the complicated logistics of logging at this time, lumber mills were built as close as possible to the timber, and the boards were brought to the site. Mills in the late 1800s Columbia Gorge were mostly small, family-run operations, analogous to family farms. (Courtesy of US Army Corps of Engineers.)

Installing the lock gates was a daunting process. The functional components that made the gates move were installed once the gates were in place. The lock gates turned through a series of counterbalances and water pressure piped from nearby Dry Creek. This photograph shows one of the massive bearings on which the gates balanced. (Courtesy of US Army Corps of Engineers.)

This photograph shows the process of getting the mixed cement mortar down to the base of the canal and laying the stone blocks for the culverts in the canal wall. Using large pipes allowed the thick layer of cement to reach the bottom layer. A layer of stone blocks was then laid on top of the cement. Earth could later be backfilled against the stone blocks to support the walls. The spaces between the blocks are for the culverts to allow water to fill and drain from the lock chambers. This image also shows the ingenious engineering of the carts. As the Cascade Locks were built, the carts were updated to include counterbalances and lever points that allowed them to tip with less force. These carts could tip in two directions, allowing the material to be loaded and dumped more easily. (Courtesy of US Army Corps of Engineers.)

Locks work by changing the water level inside a canal. A ship enters the lock chamber, and the gate closes behind it. Lock tenders turn a valve to allow water to fill or drain the chamber through culverts, large pipes along the bottom of the canal with openings every 38 feet. Water would fill the culverts from upstream to raise the water level, or drain downstream to lower it. It took 90 seconds to open or close the gates and only 10 minutes to fill or empty the lock chamber. These photographs show the installation of steel pipe and stone openings for culverts and hydraulic operation. (Both, courtesy of US Army Corps of Engineers.)

After 18 long years of difficult construction, the Cascade Locks and Canal opened. While the locks officially opened for operation on November 5, 1896, they opened with just the basic mechanics and walls fully finished. At the time, congressional appropriations and spending had reached $3,793,496.94. While it was a slow and expensive project, building the canal increased knowledge of the geology of the riverbed and techniques that worked on this geography, which influenced the design and construction of the canal at The Dalles and eventually the Bonneville Dam. Finishing touches to embankments and aesthetic details continued through 1914, as well as ongoing maintenance. This photograph shows the final product of the Cascade Locks as it was designed, with neatly stacked stones, smooth rounded edges, and a sweeping view of the Upper Cascades. (Photograph by Jesse A. Meiser, courtesy of Cascade Locks Historical Museum.)

Three

Active Use and Finishing Touches

Upon completion, the Cascade Locks were 90 feet wide, with a lower entrance width of 140 feet and an upper entrance width of 250 feet. The lock was 462 feet long and could accommodate ships to any stage up to 42 feet above low water. If necessary, the upper guard gate could function as a second lock. (Courtesy of Cascade Locks Historical Museum.)

When the Cascade Locks and Canal opened, ships could finally pass through a navigable waterway all the way from Astoria to Portland to The Dalles without relying on expensive portage. Celebrations and picnics took place at the locks, as well as in Portland and The Dalles, and in Vancouver and Camas, Washington. Hundreds of people passed through the locks on steamboats to view the final project up close and to be a part of the grand opening. The steamer *Sarah Dixon* had a small cannon on deck and blasted salutes at periodic intervals. The *Regulator*, *Dalles City*, *Maria*, *R.R. Thompson*, *Water Witch*, and *Harvest Queen* also went through the locks on opening day. As an added challenge between captains, *Sarah Dixon* and *Regulator* raced to be the first boat from the locks to The Dalles. *Sarah Dixon* won, arriving a full 30 minutes ahead of *Regulator*. (Courtesy of Cascade Locks Historical Museum.)

While the locks were fully functional when they opened, there were many finishing touches to be completed on the project for years to come. Because the locks were functional, funding them was no longer a priority, and work progressed slowly. Derricks and debris still littered the area surrounding the canal as a smaller crew of workers continued to clean up. Private contractors were hired to complete the final tasks. Slope pavement, dredging, painting, and the construction of additional staff housing continued through 1914, in addition to maintaining the structures. The photograph above shows two ships transiting the locks surrounded by support frames holding up freshly paved slopes. The image below shows continued paving work on the south slope. (Both, courtesy of Cascade Locks Historical Museum.)

The peak for passenger service and steamship travel happened before the final touches were put on the canal. In 1906, the unpolished, unfinished Cascade Locks and Canal saw a record 133,070 passengers, its highest volume in a single year. In the photograph above, the *Telephone* and *Bailey Gatzert* are fully loaded and passing through the locks while the unpaved bank sits neglected. These ships both had special four-note musical steam whistles. In the March 29, 1895, issue of the *Daily Astorian*, the whistle of *Bailey Gatzert* was described as sounding "like the bellowing of a cow just recovering from a severe attack of the grippe." It was soon repaired to become "the most musical whistle of any steamer on the river." (Both, courtesy of Cascade Locks Historical Museum.)

The photograph above was given to the museum by one of the first lock tenders, Robert G. O'Connor. It captures the view from a ship passing through the unfinished canal at its highest water level. The top of a filling culvert is just visible above the waterline next to the open gate. Below is one of the last photographs showing construction work, with the *Annew* docked against the unpaved south wall of the locks on March 29, 1914. The final payment allocated to the completion of the project was for the completion of the south wall of the upper lock on August 28, 1914. (Both, courtesy of Cascade Locks Historical Museum.)

The benefits to river commerce were immediate. In its final year of operation, the state portage railway carried 8,122 tons of freight. The canal carried many more times that at no charge to any shipper, without the need to unload and reload ships. Portage freight charges were eliminated. The Union Pacific Railroad purchased a majority stake in OR&N in 1898 and absorbed the line into its network between 1910 and 1936, becoming the new sole rail servicer on the Columbia River. By making shipping via the river more economical, the locks restored competition between the railroad and multiple river shippers. The image above shows the *Dalles City* in the locks, loaded with flour sacks. The image below shows *J.N. Teal* and *Dalles City* in the locks. *J.N. Teal* is transporting covered wagons on its bow. (Both, courtesy of Cascade Locks Historical Museum.)

In the first six months, 4,941 tons of freight and 9,081 passengers passed through the canal. That number more than tripled in 1898 to 18,812 tons of freight and 30,327 passengers. Passenger travel peaked in 1906, with 133,070 people transiting the locks. Freight tonnage varied between 16,000 and 55,000 tons per year. The value of freight through the locks in most years exceeded the cost of construction. This period saw an interesting transition in transportation technology. In the image below, *Dalles City* (left) has automobiles on its bow, while *Tahoma* (right) carries a farm wagon. (Both, courtesy of Cascade Locks Historical Museum.)

The freight carried through the locks was as varied as the agricultural areas the river and canal served. Wheat and merchandise made up two thirds of what was shipped in a typical year. Fruit, logs, livestock, lumber, salmon, and other sundries passed through the locks on their way around the world. Freight was hauled an average of 90 river miles. Log rafts up to 300 feet long and 70 feet wide could be brought through the Cascade Locks. In these photographs, *Tahoma* (above), and *Warco* (below) tow log rafts through the locks. *Warco* was owned and captained by Cascade Locks resident Karl Rosenback. The *Weown* (not pictured) was also often used to transport log rafts. (Both, courtesy of Cascade Locks Historical Museum.)

In 1891, The Dalles, Portland, and Astoria Navigation Company (DP&AN) planned a fleet of ships that would regulate the rates and fares from the longstanding monopoly enjoyed by OR&N. It commissioned four ships that would become known as the Regulator Line. This was well received by many companies frustrated with OR&Ns freight rates. The Regulator Line eventually included three of the best-known ships on the Columbia: *Regulator*, *Dalles City*, and *Bailey Gatzert*. These three made the most transits of the Cascade Locks and hauled the most passengers. (Both, courtesy of Cascade Locks Historical Museum.)

The first ship in the fleet, *Regulator*, was a $30,000, 152-foot-long stern-wheeler built by Louis Paquet in The Dalles. The *Regulator* first operated on the Middle Columbia, running freight from The Dalles to the 1891 state portage railway. Capt. Fredrick Sherman was the first captain of the *Regulator*. In 1898, *Regulator* wrecked in the Upper Cascades when it missed the approach to the locks. When the captain attempted to reverse and reapproach, the *Regulator* was blown off course by the wind. A total of 120 passengers needed to be rescued. The *Regulator* was rebuilt and ran the Columbia until it was destroyed in an explosion in 1906. (Both, courtesy of Cascade Locks Historical Museum.)

The second stern-wheeler of the Regulator Line, the *Dalles City*, was built to expand the capacity of DP&AN. The *Dalles City* was considered a twin ship to the *Regulator*, being similar in size and engines, and running the same route on alternating days. *Dalles City* carried both freight and passengers between Portland and The Dalles. The *Dalles City* and the *Regulator* ran opposite routes between Portland and The Dalles, with one traveling east one day and the other traveling west, then switching. Occasionally, they were in the Cascade Locks at the same time. (Above, courtesy of History Museum of Hood River County; below, courtesy of Cascade Locks Historical Museum.)

The trip from Portland to the Dalles became an easy day trip with navigational improvements. Scenic cruises through the Gorge increased in popularity as the journey became less difficult and ships were built for speed. Tens of thousands of passengers passed through the Cascade Locks each year. The millionth passenger transited the locks in 1912. Tourism increased, with passenger ships offering excursions for pleasure and not solely for commerce. The most popular excursion ship was the *Bailey Gatzert*, the only steamer on the Columbia River that was commemorated with a song and a postage stamp. (Both, courtesy of Cascade Locks Historical Museum.)

Bailey Gatzert was built for John Leary in the shipyard of John J. Holland at Ballard, Washington. It cost $100,000 to construct. Work was completed on October 4, 1890. *Bailey Gatzert* was 177 feet long and driven by twin horizontally mounted single-cylinder poppet valve steam engines, each with a 22-inch bore and an 84-inch stroke on the piston rod. These 1,300-horsepower engines could drive the steamer at a speed of over 20 miles per hour. A sternwheeler, *Bailey Gatzert* had an 18-foot wheel with 17 paddles. It was purchased by Columbia River and Puget Sound Navigation Company (CR&PSNC) in 1891. (Right, courtesy of Cascade Locks Historical Museum; below, courtesy of History Museum of Hood River County.)

Originally run in Puget Sound, *Bailey Gatzert* transferred to the Columbia River in 1892 and made its first excursion in the Columbia River Gorge in July 1893. After a rivalry between CR&PSNC and OR&N along the Astoria route, a $20,000 remodel, acquisition by the Regulator Line, the unionization of its deckhands, and another $25,000 remodel, *Bailey Gatzert* was reassigned to the regular route between Portland and The Dalles in May 1901. The daily run left Portland at 7:00 a.m. and embarked for the return trip from The Dalles at 5:00 p.m. (Both, courtesy of Cascade Locks Historical Museum.)

A reprint from the *Skamania Pioneer* in the February 6, 1903, *Hood River Glacier* alleges that *Bailey Gatzert* was "the first boat which demonstrated the fact that a round trip could be made between Portland and The Dalles in one day" and "also has the honor of having made the fastest time between The Dalles and Portland." (Courtesy of Cascade Locks Historical Museum.)

During the 1905 Lewis and Clark Exposition in Portland, *Bailey Gatzert* cruised through the Cascade Locks twice a day, leaving at 8:30 in the morning and 5:30 in the afternoon. The fare was $1.50 and included a special meal served on board the ship. The "Bailey Gatzert March" was composed to celebrate the occasion. (Courtesy of Cascade Locks Historical Museum.)

Bailey Gatzert was remodeled and reconstructed again in 1907. The length was extended 15 feet, adding a new freight house and removing aft cabins to include a promenade. With its new dimensions, it was licensed to carry 350 passengers regularly and 625 for excursions. Fred H. Sherman served as the first captain of the "new boat with an old name," as reported in the 1894 chief's report. (Photograph by William L. Finley, courtesy of Oregon Historical Society.)

With the renewed ship came renewed rivalry. *Bailey Gatzert* and the also recently remodeled *Charles R. Spencer* raced from Portland to Vancouver to The Dalles on May 8, 1908. *Bailey Gatzert* maintained a lead of two lengths up to the Cascade Locks and transited first, arriving at The Dalles 18 minutes before *Charles R. Spencer*. (Courtesy of Cascade Locks Historical Museum.)

After 1911, steamship travel fell out of fashion in favor of rail and car travel. Fares began to fall with reduced passengers. By the summer of 1915, only four passenger ships traveled out of Portland, with *Bailey Gatzert* providing Columbia River Gorge excursions. In 1917, *Bailey Gatzert* resumed regular runs between Portland and The Dalles and was the only ship to do so. The Cascade Locks and Canal closed for high water on June 19, 1917. This spring freshet reached 32.1 feet above low water. According to the *Morning Oregonian*, Capt. Archie Geer "treated 125 passengers to a thrill when he navigated the [*Bailey Gatzert*] through the Cascades of the Columbia" while the locks were closed. This was the first time the rapids had been run with regular passengers since 1893, and was the highest recorded water of any steamship run. In 1918, *Bailey Gatzert* was sold to the Puget Sound Navigation Company, which moved the steamer back to Puget Sound. The steamer was retired in 1926. (Courtesy of Cascade Locks Historical Museum.)

The stern-wheeler *Tahoma* was a 118-foot paddle steamship owned and operated by the first licensed female steamboat captain on the Columbia River, Capt. Minnie Mossman Hill. Captain Hill passed the master's exam in 1887 at 23 years old. She drove the ships in her company, and her husband, Charley, was the engineer. (Photograph by C.S. Reeve, courtesy of Cascade Locks Historical Museum.)

The *Tahoma* was rebuilt in 1910 and sold to former Cascade Locks resident Capt. Charles Nelson. In 1911, the White Collar Line that operated the *Tahoma* announced that it would carry automobiles to any point in the Middle Columbia that people wished to explore. Cars became common freight for the *Tahoma*. In this photograph, the *Tahoma* transits the locks on February 28, 1915. (Courtesy of Cascade Locks Historical Museum.)

The year 1916 was a dramatic one for the *Tahoma*. On January 5, the ship froze in ice for six weeks near Crown Point, Washington. Twenty passengers were rescued, and Capt. Charles Nelson, engineer Charles Kern, steward John Featherstone, mate Tom Shepherd, and a single bull waited on board. Food for the crew and hay for the bull were available via a path to Cape Horn, and the ship was transporting 100 crates of apples from Hood River. The greatest concerns were that the ship would be crushed between the ice and rocks, or that it would be released from the ice too quickly and lose control. "Boat Pilots today don't have to worry about the ice," Nelson said in commenting on the story, "The dam can control the ice by raising or lowering the level, which breaks up the ice and clears the channel." On August 1, the *Tahoma* nearly capsized after missing the entrance to the locks. The same day, it became beached on a sandbar near Viento. (Courtesy of Cascade Locks Historical Museum.)

Before dams were built, winter ice flows and surface freezing were an annual challenge on the Columbia. The river would close for a few weeks most winters to wait for conditions to improve to allow passage. On January 5, 1909, a strong east wind froze the water inside the Cascade Locks, with the *Sarah Dixon* and the *Regulator* inside. The ships were trapped for six weeks. One of the stewards from the *Sarah Dixon* was a former Swiss dancing teacher. With nowhere to go during the storm, they offered dancing lessons on board the stuck ships and in the Odd Fellow's Hall in town. Many residents walked through a foot of snow to learn to dance. Some residents took advantage of the freeze and walked across the ice to Stevenson, Washington. (Photograph by K.A. Pierce, courtesy of A.H. Warren.)

As the final touches were completed on the locks, new government employees came to manage the daily operations of the canal. While he worked on the project from 1891 to 1897, engineer and chief lock tender Valentine "Val" Tomkins brought his family permanently to the site in 1907. Tomkins was the sole office employee, and kept meticulous engineering notes, recording every detail of the goings-on at the locks. His records include regular maintenance, new additions and upgrades, the measure of every type of freight that passed through the locks, and how the weather affected daily water levels. Tomkins's notes informed the annual reports for the Corps of Engineers. He was also responsible for high-level maintenance, completing work from hiking up the access trail along Dry Creek to inspect the pipeworks at the falls, to donning a diving suit and performing underwater maintenance on the equipment. (Courtesy of Dr. Alan Tomkins.)

This building was erected by Ball and Platt in 1874 and served as an office and cement testing facility through construction. It became the engineering and lock tender's office after construction was complete. Val Tomkins worked in this office for 27 years, completing analysis on water samples, testing cement, and recording data. (Courtesy of Cascade Locks Historical Museum.)

Three additional lock tenders were hired to operate the locks. The lock tenders raised and lowered the locks an average of five times per day and performed maintenance on the many systems that kept the locks operating. The lock tenders lived in government housing facing the locks so they could have easy access when ships arrived. (Photograph by K.A. Pierce, courtesy of Cascade Locks Historical Museum.)

Lock tenders at Cascade Locks lived on-site with their families. The lock tender quarters were functional foursquare houses, with front doors facing the river. Lock tender Robert G. O'Connor and his family moved into the lock tender quarters on June 11, 1907. His primary job was to manually open and close the locks when ships arrived. This family photograph on the porch of the house shows, from left to right, (first row) Ruth O'Connor, Robert G. O'Connor Jr., and Edna Washburn; (second row) Robert G. O'Connor Sr., Bertha O'Connor, Hope O'Connor, and Emma Jensma. Lock tenders had many other duties. Below, O'Connor stands in the maintenance shed with a model of a fish wheel he built for the 1905 Lewis and Clark Exposition. (Both photographs by K.A. Pierce, courtesy of Cascade Locks Historical Museum.)

Lock tenders' families had unlimited passes to ride ships to and from the Cascade Locks. Frequent trips were made to Hood River for church, school, and shopping. Pictured here from left to right are Mabel Adams, Bertha O'Connor, Ella States, and Sophie O'Bryan. They are drinking lemon juice through peppermint sticks aboard the *George W. Simon*, captained by Capt. Charles Nelson. (Courtesy of Cascade Locks Historical Museum.)

Lock tending was a job that required immediate availability. When a ship arrived, it needed to be lifted or lowered speedily. Lock tenders lived on-site with their families. Several families lived on the government reservation over the years of the operations of the Cascade Locks and Canal. (Courtesy of Cascade Locks Historical Museum.)

The valve house for the Cascade Locks housed the turbines, pumps, and connections to the water lines that powered the hydraulic system. Water would enter the powerhouse through the pipes from the sources at Dry Creek and Moody Avenue Reservoir and turn the turbines and pistons to move the gate valve. The pressure would open the gates or the culvert valves, allowing the water to flow and ships to move. The photograph above shows lock tender Brolliar tipping his hat to a passing steamship in 1917 or 1918. The photograph below shows where the valve house that he is standing in front of was located in relation to the lock gates. (Both, courtesy of Cascade Locks Historical Museum.)

The lock gates were painted with red lead paint. Tomkins recorded the recipe for six and a half quarts of this paint. It called for 35 pounds of dry red lead, one gallon of linseed oil, one half-pint of turpentine, and one half-pint of brown Japan dryer. This would cover 99.5 square feet of the lock gates, with two coats on the lock gates requiring 22,285 square feet of paint, or 224 times the recipe. The lock gates were repainted as part of regular maintenance. Photographs were frequently colorized for postcards. Most of the colorized postcards from the Columbia River were colored in Germany, by people who had never seen the colors as they really were. This led to widely distributed images of the locks showing steel gray gates. (Courtesy of Cascade Locks Historical Museum.)

Massive amounts of water were used to operate the locks. A total of 3,360 gallons displaced the pistons to fill the lift chambers in 10 minutes to raise the ships. Only 840 gallons were actually needed to displace the pistons; the rest was allowed for waste as water flowed through wooden six-inch pipes and old steel five-inch pipes. These pipes could carry 790 gallons per minute. The pipelines were eventually replaced with 10-inch steel pipe. The photograph above by Lily E. White shows the gates midway through opening. The dark line in the photograph below shows the high water mark inside the locks. (Above, courtesy of Oregon Historical Society; below, courtesy of Cascade Locks Historical Museum.)

While passenger traffic declined with the popularity of car travel and new access via the Columbia River Scenic Highway, freight traffic steadily increased as industry boomed. Wheat farming and cattle ranching in the high desert regions of Eastern Oregon and Washington and a booming logging industry kept consistent river traffic flowing through the locks, and growing communities required merchandise to be sent upriver from the ports at Portland. In the 10 years leading up to their closure, $6,418,891 in freight traveled through the locks, including an average of 50,000 tons of logs rafted through the canal. (Both, courtesy of Cascade Locks Historical Museum.)

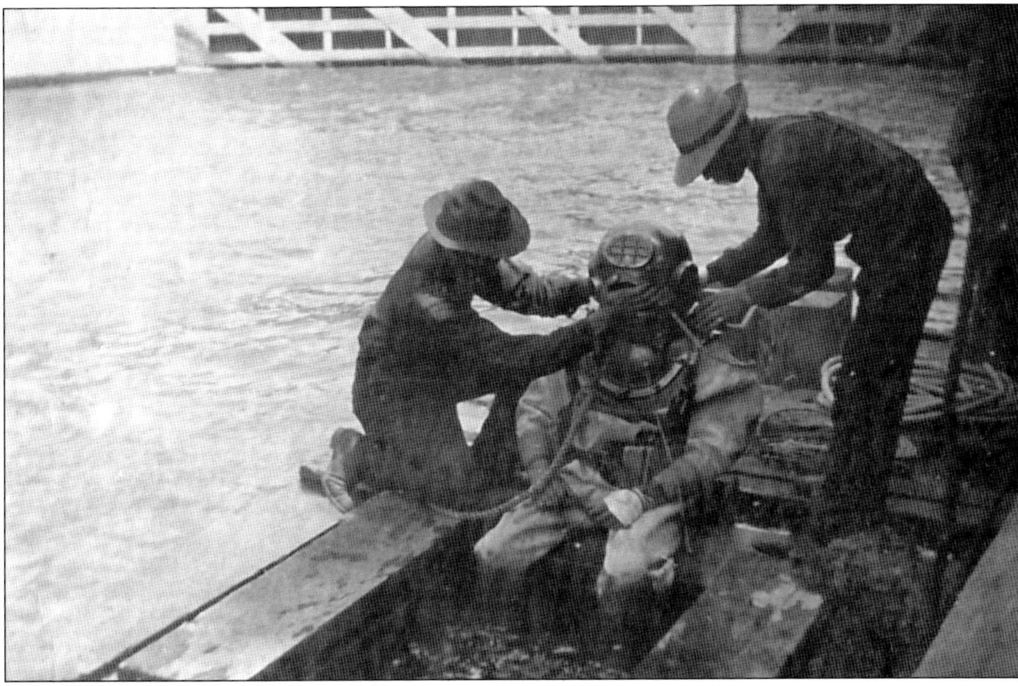

With so much of the mechanics happening underwater, sometimes it was more practical to send the lock tenders down to inspect valves and culverts than it was to drain the canal. In these photographs, lock tender O'Connor dons a diving suit to perform maintenance on a valve. The design of the locks left an open pit for maintenance access to the culverts and valves, so maintenance diving frequently took place in these enclosed spaces instead of out in the lock chamber. The diving suits had lead-filled boots to help with descent and plate-glass faceplates for underwater viewing. (Both, courtesy of Cascade Locks Historical Museum.)

In 1902, the Yacoult Burn blazed through the Columbia Gorge, burning from Bridal Veil to Eagle Creek on the Oregon side and between the Lewis River, Wind River, and Washougal River on the Washington side. Additional burns occurred in 1919, 1922, 1927, and 1929. Here, damaged trees overlook the complete Cascade Locks. The ship in the locks is a dredge. The canal required dredging after the annual spring freshet. The high-water crest was typically in early June, and

dredging would take place in late summer. The steamer *Cascades* was a government-owned bucket dredge that would perform this task, frequently assisted by hired scows. An average of 25,000 tons of material would be removed from the canal and its entrances. (Courtesy of Cascade Locks Historical Museum.)

Running ships upriver against the current was a little easier at low water. These scenic photographs of the *Regulator* approaching and the *Tahoma* leaving the Cascade Locks at low water show just how high the water level came on the river banks at the Upper Cascades. The exposed rocks and gravel mark where the high-water line hit that year, and the flat area on top of the embankment is the Union Pacific railroad line. The photograph site was a high point above the railroad line, which would later become part of the Columbia River Highway and the southern terminus for the Bridge of the Gods, built in 1926. (Both, courtesy of Cascade Locks Historical Museum.)

With the stonework finally completed in 1914, trees grown in, and picturesque foursquare lock tender houses built, the Cascade Locks became a destination for photographers looking to capture the beauty of the region and the ingenuity of the engineering of the structure. The worker barracks were relocated to the main road to become the Columbia Hotel, where they remained until the Cascade Locks urban renewal project in 1972. Moving the barracks created a clear line of sight between the three lock tender houses. As car travel increased, scenic overlooks were built to provide ample opportunities for photography. The overlook below on the Washington side was raised after construction of the Bonneville Dam. (Both, courtesy of Cascade Locks Historical Museum.)

As car traffic increased on the scenic highways, demand grew for a bridge between Oregon and Washington. In 1920, the Corps of Engineers contracted Interstate Construction Corporation to build a bridge. By 1925, the company had only constructed one pier. Wauna Toll Bridge Company purchased the bridge and completed construction in 1926. The cantilevered steel bridge had a wooden deck 91 feet above the river. The bridge was raised 44 feet in preparation for the completion of the Bonneville Dam. In 1953, the Columbia River Bridge Company acquired the bridge, and the Port of Cascade Locks purchased it on November 1, 1961. In 1967, the bridge was re-decked in steel. The photograph above shows the bridge at its original height. Below is the bridge as it appears today. (Both, courtesy of Cascade Locks Historical Museum; below, photograph by Janice Crane.)

The scenic overlooks were very popular with photographers, boat and train enthusiasts, and general travelers. As photography developed and became more affordable and portable for the recreational user, more photographs were taken of the locks, canal, and Cascades. Photographs were also sold as souvenirs for travelers. As a result, there are many photographs that appear similar, all taken from the angles of the primary scenic overlooks and lower-traffic, wider viewpoints of the locks. Common subjects include the locks and Cascades, late-running passenger ships like *Bailey Gatzert*, and later-running freight ships. (Above, courtesy of Oregon Historical Society; below, courtesy of Cascade Locks Historical Museum.)

As a way to gain notoriety and press for the shipping companies and also as a display of skill of the pilots, stern-wheeler racing was relatively common. In the photograph above, the *Charles R. Spencer* approaches the locks just a few lengths ahead of the *Bailey Gatzert*. The most well-publicized race between the two was on May 8, 1908, when they both had fresh remodels. Below, the *Bailey Gatzert* approaches the locks alone. (Above, courtesy of US Army Corps of Engineers; below, courtesy of Cascade Locks Historical Museum.)

Steamships relied on heat to run their engines. The first steamships on the Columbia River burned wood fuel. Some of the wood-burning stern-wheelers consumed massive amounts of fuel, burning an average of four cords of wood per hour. When settlers along the Columbia found the rocky volcanic soil unsuitable for farming, many set up small logging operations and operated fuel stations to supply ships with wood. As technology developed, some ships were retrofitted to burn oil. Oil-burning steamships had to stop less frequently for fuel, but were more prone to explosions. In historic photographs, wood smoke appears white (above), and oil smoke looks black (below). (Both, courtesy of Cascade Locks Historical Museum; below, photograph by K.A. Pierce.)

Steamships were not the only boats on the Columbia River. In this photograph, fishermen in a small craft paddle out of the way as the *Charles R. Spencer* approaches the lower entrance to the locks. The fish wheels rapidly depleted the number of fish in the Upper Cascades. (Photograph by K.A. Pierce, courtesy of Cascade Locks Historical Museum.)

The *Georgie Burton* was one of the last steam-powered ships on the Columbia River, making her final trip from Portland to The Dalles on March 20, 1947. The ship was named for Georgiana Burton Pittock, a famed pioneer, community leader, suffragette, and philanthropist in Portland. Pittock Mansion is open to the public today. (Courtesy of Cascade Locks Historical Museum.)

The gates of the Cascade Locks also functioned as a pedestrian bridge. They were roped and chained across the top to make crossing them safer. Ships could moor on the north side of the canal and still allow passengers to access the mainland. (Courtesy of Cascade Locks Historical Museum.)

This photograph shows the *J.N. Teal* heading out of the locks and up the river. In 1915, the *J.N. Teal* had the honor of being the first ship to pass through the Celilo Locks and Canal upriver at The Dalles, making it the first ship to transfer between the Middle and Upper Columbia. (Courtesy of Cascade Locks Historical Museum.)

This series of photographs demonstrates the change in water level inside the locks. Above, the water is at a high level, filling over the lowest and middle staircases. Below, the water is at a much lower point, with the pavement and lowest staircase visible. The churn in the canal is the culvert draining and lowering the ship in the lock. On the opposite page, the ship in the locks is barely visible as the locks are drained to a lower level. The water line on the canal wall shows how full the lock was. (Both, courtesy of Cascade Locks Historical Museum.)

A cow grazes in the background at right side in this photograph. Federal employees followed strict rules regarding animals on site. Alice Tompkins Fee recalled that "Uncle Sam allowed us to have twelve chickens, and . . . a cow. But we couldn't 'compete with the villagers.' . . . We couldn't sell eggs, and we had an awful time getting rid of the milk when the cow was fresh." (Courtesy of Cascade Locks Historical Museum.)

The section of the Columbia River Highway from Portland to Hood River was completed in 1916, with the highway reaching the Cascade Locks in 1915. Many preferred the novelty and potential for spontaneity of travel by car. The Cascade Locks and Canal became a popular place for motorists to stop and picnic on their excursions. (Courtesy of Cascade Locks Historical Museum.)

The Cascade Locks and Canal are in a notoriously interesting weather area. While the canal walls sit at just 105 feet in elevation, the community is nestled at the base of the dramatic 4,000-foot-tall mountains that make up the walls of the Columbia River Gorge. The far eastern reach of the tides and marine clouds from the Pacific Ocean strike the crest of the Cascade Mountains, creating the rain shadow that leaves arid lands to the east. Spring rains bring landslides and floods, followed by a drying out for the summer that ends in a roaring wildfire season. Winter brings piles of mountain snow. The east winter winds gust up to 90 miles per hour, enough to freeze the Columbia River. (Both, courtesy of Cascade Locks Historical Museum.)

Four

Decommission and Remnants

Federal responses to the Great Depression brought more public works projects. As part of his 1932 presidential campaign, Franklin Delano Roosevelt promised to harness the hydropower of the Columbia River. On February 9, 1938, the spill gates of Bonneville Dam closed and the river inundated the Cascade Locks. The Cascade Locks and Canal project was officially abandoned by June 30, 1939. (Courtesy of Cascade Locks Historical Museum.)

Bonneville Dam was the first federal dam and the second of 60 dams built on the Columbia River and its tributaries. As the first dam in the area, construction had to contend with rapid changes in water levels. A great deal of relocation work was required to accommodate the rising levels. The Bridge of the Gods was raised 44 feet to accommodate the new water level. Several miles of Washington Highway 8, Oregon Highway 30, railroads on both shores, and the fish hatchery at Bonneville had to be relocated above the new water level. A total of 2,500 acres of land were purchased and cleared along the river at the reservoir site. The new reservoir flooded Native American tribes' treaty fishing sites, and new sites had to be negotiated. This photograph shows the new water level and inundated locks from the new site of the scenic overlook on Washington Highway 8. The buoy in the water marks where the lower entrance to the canal was located. (Courtesy of Cascade Locks Historical Museum.)

Bonneville Dam is just upriver from the middle landing of the old portage railroad, near the mouth of Eagle Creek and at the lower end of the river improvements that were completed for the approach to the Cascade Locks. This is also the easternmost point of the Columbia River to be affected by the Pacific Ocean tides. Ships traveling upriver still needed a lock to access the new reservoir. The navigation lock at Bonneville fulfilled this function. The first powerhouse at Bonneville Dam originally had 10 generators, the first two installed by 1938, and the rest by 1944, generating 43,200 kilowatts of power each. When hydroelectric power came online from Bonneville Dam, the first line was run three miles upriver to Cascade Locks. Before the Bonneville Power Association, most people in Cascade Locks did not use electricity, instead using kerosene and coal to cook and heat their homes. The power lines are visible next to the Union Pacific Railroad tracks in this photograph. (Courtesy of Cascade Locks Historical Museum.)

While the planned water level would rise significantly in the reservoir above the dam, additional river improvements were needed at the Upper Cascades, Sheridan Point, and Three Mile Rapids to create a deep-water channel for ships to pass. The preparation for the channel around the outside of the Cascade Locks required blasting the remaining large landslide detritus to finally eliminate the Upper Cascades as a barrier to water transportation. As technology improved, ships were built bigger. The navigation lock at Bonneville was built with a deep draft to allow for these bigger, more heavily loaded ships to pass, and the improved channels needed to be deep enough to accommodate them. As with earlier river improvements, dynamite was the tool of choice for efficiently removing large quantities of rock and sediment. This photograph shows the blasting of some of the large boulders in the Upper Cascades. (Courtesy of US Army Corps of Engineers.)

These photographs show before and after the blast on the previous page. Above, workers dig trenches and set dynamite in the boulders on the north side of the canal. Setting dynamite was precise work; it needed to be placed in the right spots on a boulder to send the debris away from the workers and out into the water. The photograph below shows the aftermath and the rubbly remains of the boulder after the blast. The smaller pieces were removed by laborers using carts. (Both, courtesy of US Army Corps of Engineers.)

Disastrous flooding struck again from May 23 to June 5, 1948. An unseasonably cool early spring caused ice and snow to stick around the Columbia River in British Columbia and eastern Washington. When two very large May rains finally arrived, the snow and ice melted quickly, and most of the tributaries of the Columbia reached flood stage. Stream gauges recorded near–record high water discharge rates, most only surpassed by the 1894 flood. The photograph at left shows water pouring out of the lower end of the canal. Below is a closer view of the lower end with a portion of the damaged lock gate. (Both photographs by Pat Dressler, courtesy of Cascade Locks Historical Museum.)

On May 28, 1948, engineers at the Bonneville Dam reported that 680,000 cubic feet of water per second was passing through the spillways of the dam. This was about 95,000 cubic feet above average for high water. The 1948 flood became known as the Vanport Flood because it destroyed the town of that name, the second largest and most racially diverse city in Oregon. The flood displaced over 18,500 residents at Vanport, one-third of whom made up the largest Black community in Oregon. The photograph at right shows water dropping into the Cascade canal nearly level with the upper stonework. Below, water pours through the canal, with the Bridge of the Gods in the background. (Both photographs by Pat Dressler, courtesy of Cascade Locks Historical Museum.)

The water from the tributaries poured into the Columbia River and eventually to the Cascade Locks and Canal. Highway 30 closed on May 28 for flooding between The Dalles and Cascade Locks. The Dalles waterfront was inundated and the canal was closed. This photograph shows water entering the upper entrance of the locks and how high the water level came up the stairs. (Photograph by Pat Dressler, courtesy of Cascade Locks Historical Museum.)

The pressure from the flood was so great that it dislocated and bent one of the steel lock gates. This photograph shows how the gate rotated out of position and curled along the canal wall, along with the snapped cable that would have been part of the hydraulic control system. (Photograph by Pat Dressler, courtesy of Cascade Locks Historical Museum.)

In 1953, the Corps of Engineers removed the upper lock gate and guard gate. The lower lock gate had been underwater for 15 years and was deemed too silted-in to be removed and was left in place. This photograph shows the derrick barge *Cascade* lifting the lock gate out of position. (Photograph by G.R. Hesgard, courtesy of Cascade Locks Historical Museum.)

The gates were then placed on the ground adjacent to the canal. They were dismantled, removing the rivets to break them into pieces, then hauled away. The lock gate steel was melted down to be used in other projects. Note the size of the gate compared to the man and pickup truck next to it. (Photograph by G.R. Hesgard, courtesy of Cascade Locks Historical Museum.)

The Corps of Engineers remitted the remnants of the Cascade Locks and Canal to the Port of Cascade Locks in 1954. As required by the agreement, the port began to develop the property for recreation, creating Cascade Locks Marine Park. The port undertook major changes to develop the property and continues to make improvements to the park. A boat ramp was installed over the embankment that led to the now flooded lower canal. Asphalt was poured on the south side of the canal for parking. In 1970, an RV campground was installed on the site of the relocated worker barracks. The original visitor center for the Bonneville Dam was barged upriver and placed next to a new marina, and is now the café where the excursion stern-wheeler *Columbia Gorge* offers day trips in the summer. This photograph shows people fishing on the old canal wall in August 1959. The canal remnants remain a popular fishing spot. (Courtesy of Cascade Locks Historical Museum.)

In 1969, the Port of Cascade Locks installed a footbridge where the upper lock gate formerly allowed pedestrians to cross the canal. Residents recall that the workers had to let the air out of the tires of the truck to unload the new bridge at the correct level. This new bridge allowed recreational access to the north side of the canal, now dubbed "Thunder Island." The name was inspired by the memory of the roaring sound of the Cascade Rapids, which could be heard throughout the town prior to the construction of Bonneville Dam. Gene Miller, mayor of Cascade Locks, led the first group to cross the bridge. (Both, courtesy of Port of Cascade Locks.)

In 1969, a group of devoted community historians worked with the City of Cascade Locks to form a museum commission. The museum would preserve local history and bolster tourism after the completion of Interstate 84 re-routed traffic around downtown. The Port of Cascade Locks leased the first of the three remaining lock tender houses to become the Cascade Locks Historical Museum. The former Quarters Two was now dubbed House One, and filled with the belongings of several of the earliest settler families to arrive in the area. In 1970, the Port of Cascade Locks worked with the Oregon Historical Society to return the *Oregon Pony* to Cascade Locks. (Above, courtesy of Cascade Locks Historical Museum; below, courtesy of Port of Cascade Locks.)

The *Oregon Pony* was donated to the State of Oregon after the 1905 Lewis and Clark Exposition. The state collection is managed by the Oregon Historical Society (OHS). From 1905 to 1929, the *Pony* was stored in a Union Pacific warehouse for OHS. In 1929, it was repainted and put on display in front of Union Station in Portland. In 1969, the Port of Cascade Locks petitioned the board of OHS to return the locomotive to Cascade Locks. The port constructed an open-air shelter for it. Proud of the return of the *Oregon Pony*, the Port of Cascade Locks put special effort into incorporating it into its long-term vision for Cascade Locks Marine Park. (Both, courtesy of Cascade Locks Historical Museum.)

A special shed was constructed to house the *Oregon Pony* on the lawn in front of the new museum. In the photograph above, a maintenance crew from the Port of Cascade Locks lays sod with hand tools around the *Pony* shed in front of the museum. From left to right are Donnie Latimer, Ken Hutton, Glenn Latimer, and unidentified. Below is the completed lawn and new home of the *Oregon Pony*. (Both, courtesy of Port of Cascade Locks.)

In 1970, the State of Oregon conducted a survey of historic sites and buildings. After the survey was completed, the Port of Cascade Locks worked with the Oregon State Highway Division to submit a nomination for the Cascade Locks and Canal to the National Register of Historic Places in 1974. The nomination included the defunct locks and canal structure, remains of the canal, and three standing lock tender houses. The nomination was considered "borderline" by the review panel because so many changes had been made. The National Register of Historic Places was created in 1966, long after the most dramatic changes to the locks and canal had occurred. The Cascade Locks and Canal were listed in the National Register of Historic Places on June 4, 1974, with the committee emphasizing the need to restore the 1904 lock tender houses. (Courtesy of Port of Cascade Locks.)

In 1974, construction began to add a second powerhouse to the Bonneville Dam. The new construction would raise the water level. To preserve access to the now-popular recreational uses of the old canal structure, the Port of Cascade Locks worked with the Corps of Engineers to add an additional layer of concrete blocks to raise the top of the canal five feet. The previously flooded area to the left of the canal wall in the photograph above was filled at this time to allow for the construction of a boat ramp and parking lot, while the flooded area on the island was left to pool. (Both, courtesy of Port of Cascade Locks.)

Concrete forms were built to allow the historic masonry stairs to meet with the new concrete walkways. The concrete forms had to be customized to retain the spirit of the historic structure and to ensure they could sustain use without causing further damage to the historic structure. The banisters of the stairs were broken to shorten them, bringing the stylized rounded bottom blocks up to finish the shortened staircase. Above, workers build a wooden form for pouring concrete. One is balanced on a board in the water to access the canal side of the form. Below, workers place rebar to reinforce the new concrete structure. (Both, courtesy of Port of Cascade Locks.)

A crane mounted on a barge placed blocks of concrete that were fifteen feet long, seven feet wide, and five feet tall on top of the masonry to raise the height of the canal. This new style of concrete block was designed to interlock, securing the blocks using only their own weight. The photograph at right shows the height difference between the old canal walls and the additional concrete blocks. Below, a family watches the spectacle while fishing. They were among the last people to set foot on the original masonry. (Both, courtesy of Port of Cascade Locks.)

Changes to the canal were completed in 1978. Construction of the second powerhouse at Bonneville Dam was not completed until 1981. The photograph at left shows the complete new concrete form, matched with the prefabricated concrete blocks at the final level of the canal. It also shows the newly replaced stair banister stones, flush with the new shorter staircase. The photograph below shows the concrete blocks staged in the filled parking area to be laid upon the old masonry. The parking area was filled and brought up to the level of the new height, leaving a six-inch curb for safety. (Both, courtesy of Port of Cascade Locks.)

In the late 1960s, the lock tender quarters were prepared to become the Cascade Locks Historical Museum. A devoted crew of volunteers with deep roots in the community collected their family pioneer belongings, antique furniture, quilts and Jacquard weavings, books, lock and stern-wheeler memorabilia, Wind River Lumber Company equipment, Native American arrowheads and tools found nearby, home goods, stoves, dolls, photographs, and other objects to represent the history of the community. They laid out the lock tender dwelling as a house museum, displaying objects in the rooms they would have been used in. The parlor was stocked with books, clocks, furniture, weavings, and a coal-burning stove. (Both, courtesy of Cascade Locks Historical Museum.)

The Cascade Locks Historical Museum founders paid special attention to quilts in their collecting and displays. The log cabin quilt on the headboard, fan pattern quilt on the wall, and carpenter's wheel star quilt on the bed span a design period of the 1860s to 1940s. Quilts are dated based on their newest addition, and the founders repaired many of the quilts in the collection. Mannequins are used throughout the museum. (Courtesy of Cascade Locks Historical Museum.)

Recreational cruising remains a leisure activity on the Columbia River. This photograph shows a passenger cruise ship docked in the lock remnants after the installation of the footbridge and the raising of the canal walls. Columbia River cruise ships continue to occasionally use the Cascades Canal as a docking area. (Courtesy of Cascade Locks Historical Museum.)

In 1980, the OHS determined that the *Oregon Pony* should be fully restored in honor of its 120th anniversary. After 70 years of exposure to the Pacific Northwest elements, the *Pony* was in poor repair, with lots of rust and corrosion. The OHS undertook a several-month-long study of its history and construction to ensure it was restored correctly. Gales Creek Enterprises, the operator of Oregon Trolley Park Museum, later the Antique Powerland Museum, was selected to repair the *Pony*. Care was taken to retain as much original material as possible. A new structure protects the *Oregon Pony* from the elements. (Both, courtesy of Cascade Locks Historical Museum.)

The *Oregon Pony* was repainted to match its appearance during the 1905 Lewis and Clark Exposition. While the engine is not fully functional, it is intact, and the *Pony* is able to roll on its wheels and pistons. Once fully restored in 1981, the Port of Cascade Locks enclosed and upgraded its shed. The upgrades included full climate control and ongoing monitoring to preserve the locomotive. The *Oregon Pony* is still on display in its new home in Cascade Locks Marine Park and will remain there for the foreseeable future. (Both photographs by Janice Crane, courtesy of Cascade Locks Historical Museum.)

Bibliography

Annual reports of the chief of engineers, United States Army, to the secretary of war, for the years 1876, 1878, 1883, 1894, 1896, 1938, and appendices.

Attwell, Jim. *Columbia River Gorge History, Volume Two*. Cascade Locks, OR: Tahlkie Books, 1975.

Donaldson, Ivan J. and Frederick K. Cramer. *Fish wheels of the Columbia*. Portland, OR: Binfords & Mort Publishers, 1971.

Healy, Don and Carol Craig. "Yakama Nation History." www.yakama.com/about/.

"Interview: Alice Tomkins Fee: Growing Up on the Cascade Locks 'Reservation.' " *Oregon Historical Quarterly*. Fall 1987, Volume 88, No. 3: 285–307.

Reynolds, Nathaniel D. "Dating the Bonneville Landslide with Lichenometry." *Washington Geology*. Vol 29, No. 3/4. December 2001:11–16.

Terry, Jeff. "Legacy of the Oregon Pony." *The Trainmaster*: Oregon Pony Special Issue, 2009. www.pnwc-nrhs.org/trainmaster_special_editions/Oregon_Pony.pdf.

Willingham, William F. "Engineering the Cascades Canal and Locks, 1876–1896." *Oregon Historical Quarterly*. Fall 1987, Volume 88, No. 2: 229–258.